Cambridge Elements ≡

Elements of Paleontology

INTEGRATING MACROSTRAT AND ROCKD INTO UNDERGRADUATE EARTH SCIENCE TEACHING

Phoebe A. Cohen
Williams College

Rowan Lockwood
College of William and Mary

Shanan Peters
University of Wisconsin Madison

CAMBRIDGE
UNIVERSITY PRESS

University Printing House, Cambridge CB2 8BS, United Kingdom

One Liberty Plaza, 20th Floor, New York, NY 10006, USA

477 Williamstown Road, Port Melbourne, VIC 3207, Australia

314–321, 3rd Floor, Plot 3, Splendor Forum, Jasola District Centre, New Delhi – 110025, India

79 Anson Road, #06–04/06, Singapore 079906

Cambridge University Press is part of the University of Cambridge.

It furthers the University's mission by disseminating knowledge in the pursuit of education, learning, and research at the highest international levels of excellence.

www.cambridge.org
Information on this title: www.cambridge.org/9781108717854
DOI: 10.1017/9781108681445

© The Paleontological Society 2018

First published 2018

A catalogue record for this publication is available from the British Library.

ISBN 978-1-108-71785-4 Paperback
ISSN 2517-780X (online)
ISSN 2517-7796 (print)

Integrating Macrostrat and Rockd into Undergraduate Earth Science Teaching

Elements of Paleontology

DOI: 10.1017/9781108681445
First published online: October 2018

Phoebe A. Cohen
Williams College

Rowan Lockwood
College of William and Mary

Shanan Peters
University of Wisconsin Madison

Abstract: New online resources are opening doors for education and outreach in the earth sciences. One of the most innovative online earth science portals is Macrostrat and its mobile client Rockd – an interface that combines geolocated geological maps with stratigraphic information, lithological data, and crowd-sourced images and descriptions of outcrops. These tools provide a unique educational opportunity for students to interact with primary geological data, create connections between local outcrops and global patterns, and make new field observations. Rockd incorporates an aspect of social media to its platform, which creates a sense of community for users. Here we outline these resources, give instructions on how to use them, and provide examples of how to integrate these resources into a variety of paleontology and earth science courses.

Keywords: geoscience education, macrostrat, rockd, mobile learning

ISBNs: 9781108717854 (PB), 9781108681445 (OC)
ISSNs: 2517-780X (online), 2517-7796 (print)

Contents

1 Introduction

1.1 What Are Macrostrat and Rockd?

Macrostrat (macrostrat.org) is a platform for the aggregation and dissemination of geological data relevant to the spatial and temporal distribution of sedimentary, igneous, and metamorphic rocks (Peters et al., 2018). Interactive applications built upon Macrostrat data and infrastructure are designed for educational and research purposes. Several applications that use Macrostrat-sourced data now exist, including the third-party Flyover Country app (fc.umn.edu) and Rockd, developed by Shanan Peters and the Macrostrat group at the University of Wisconsin Madison. Rockd is a mobile app and complementary web-based portal designed to facilitate the collection and use of field-based geological observations of all types. Rockd allows users to post geolocated text and image content to a database that is linked to a wide range of geological entities, including bedrock map units, lithostratigraphic rock units, lithologies, fossil taxa, and minerals.

The Macrostrat database's web application allows users to access information on both surficial geology (bedrock) and stratigraphic (column) data (Fig. 1). The bedrock view allows users to examine a dynamic global geological map of surface units, search for specific locations and lithologies, and pull up additional information on map units (Fig. 1B,D). The column view allows users to examine regional geology at a stratigraphic level (Fig. 1C). Column data are most comprehensive for North America, but additional regions such as New Zealand are now available as well. Clicking on a column polygon reveals the stratigraphy of the area, links to Paleobiology Database collections, and shows information on lithology types, literature references, and more. Macrostrat allows users to search for a particular lithology type, stratigraphic unit, or time period, and allows merged searches (e.g., searches for all columns that contain Cambrian carbonates). The broader Macrostrat dataset has been used extensively for research purposes (i.e., Finnegan et al., 2012; Peters and Gaines, 2012; Nelsen et al., 2016; Husson and Peters, 2017); however, this is the first publication to focus on educational applications.

Rockd is a mobile app based on the Macrostrat dataset. Although the app can be used to explore the global bedrock map interface of Macrostrat, its primary utility lies in the use of geolocation technology. By simply opening the app, a user is presented with Macrostrat's "best guess" for the geological unit the user is currently standing on, its age, lithology, and, for rocks younger than 600 million years, the reconstructed paleogeography for that location at the time the rock formed (Fig. 2). In addition, the map interface shows locations, most with photos that other users have uploaded of nearby outcrops (Fig. 3).

A   Macrostrat Landing Page with search optoins

B   Bedrock view with filters applied

C   Column view with filters applied

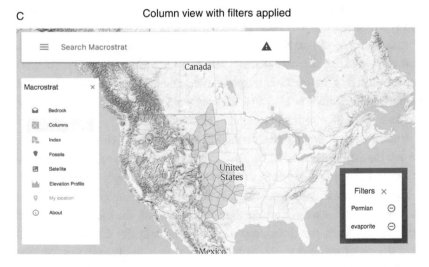

Figure 1. Macrostrat overview. A, Main Macrostrat landing page showing the search bar with options. B, Macrostrat bedrock view showing surface

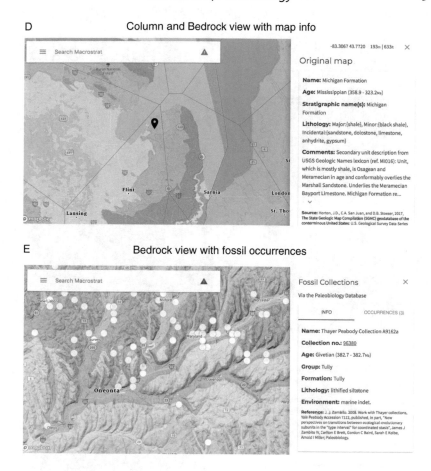

D   Column and Bedrock view with map info

E   Bedrock view with fossil occurrences

Caption for Figure 1.   (cont.)

geological units. Left inset shows layer options that can be viewed by clicking on the three horizontal bars in the search box. Right inset shows that a timescale filter has been applied, so only Devonian-aged rocks are displayed. Clicking on the small triangle in the search box displays applied filters. C, Macrostrat column view with two filters applied. Each polygon is clickable to reveal information about that geological column. D, Macrostrat column and bedrock filters. Sidebar shows information on the map unit represented by the place-marker pin; scrolling down on sidebar will also show information on column stratigraphy. E, Macrostrat bedrock view with Paleobiology Database collections layer added. Clicking on an individual circle will pull up a sidebar with data on that specific PBDB collection.

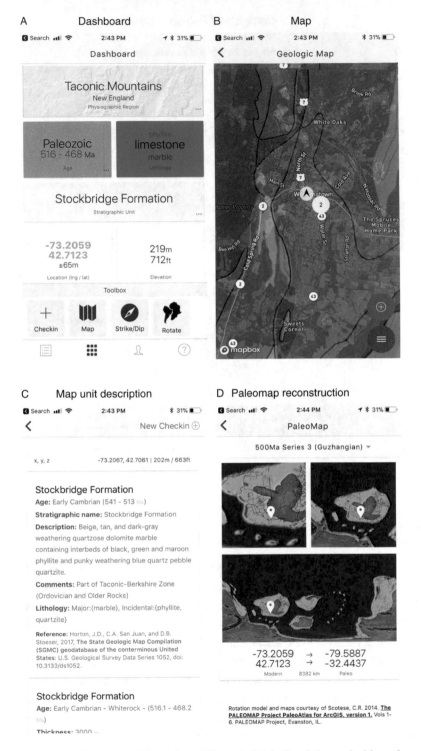

Figure 2. Example of Rockd interface. A, Main Rockd app dashboard, showing geological detail of user's current location. B, Dynamic geological

Rockd also has a complementary web-based portal (Fig. 4), without geolocation capabilities, that allows users to browse and search checkins for specific stratigraphic units, lithologies, and time periods.

1.2 Why Use These Resources in the Undergraduate Classroom?

Scientific data are increasingly moving into the digital realm, and new technologies enable better access to these data by both professionals and the general public. At the same time, STEM (Science, Technology, Engineering and Mathematics) pedagogical research strongly supports the use of primary data in classroom settings (e.g., Styers, 2018). Thus, digital repositories of scientific data are an exciting and accessible way to bring authentic science into the classroom. However, significant barriers can exist in terms of data access, the challenges of using available analytical tools, and instructor familiarity. Our goal here is to showcase how Macrostrat and Rockd can be used in an undergraduate educational setting, thus reducing barriers for course instructors and enabling more students to interact with primary geological datasets.

1.2.1 Providing the Context for Fossil Data

Both Macrostrat and Rockd integrate fossil data, either from the Paleobiology Database or from individual observations in the field. Both also explicitly focus on the temporal and spatial distribution of rocks, and thus are valuable as tools to help students understand the sedimentary record and the fossils that sedimentary rocks contain. In addition, conceptualizing the distribution of different types of rocks through space and time is a key way to understand how environments, climate, and tectonics have changed through time – factors that play critical roles in the patterns of biotic change documented throughout the Phanerozoic.

1.2.2 Mobile Learning

Mobile learning is becoming a more and more integral part of teaching and learning, especially in disciplines with place-based aspects, such as the geosciences. Examples of research on mobile learning using smart phones and other

---

Caption for Figure 2.   (cont.)

map of user's location. C, Geological map unit information available when user clicks on a map unit on the map screen. D, Paleocontinental reconstruction of user's location at time of bedrock deposition/formation.

devices show that they can add an important and welcome dimension to student learning experiences (McGreen et al., 2005; Elkins, 2009; Taleb and Sohrabi, 2012; Gikas and Grant, 2013; Dennen and Hao, 2014). Rockd is the most sophisticated mobile learning tool currently available for the geosciences. In addition to allowing students access to geological data, Rockd connects students to an international community of both professional and amateur earth scientists. This is a compelling way to empower students, as they can create their own accounts, make their own checkins, and contribute to a crowd-sourced database of geological knowledge. These attributes make Rockd an example of effective mobile learning, characterized by Gikas and Grant (2013) as learning that (a) engages learners with constant connectivity, (b) fosters collaboration, and (c) enables authentic learning on the move. One of the additional powerful aspects of Rockd is that it allows for informal learning to occur outside of class time. Once students become familiar with the app, they can use it at any time – during breaks, on vacation, on field trips for other courses, or simply during their day-to-day lives. This bridges the gap between the classroom and the "real world" in a powerful way. Thus, Rockd can serve as a link between formal and informal learning environments (Gikas and Grant, 2013).

1.2.3 Field, Research, and Active Learning Experiences Increase Retention

Research indicates that field and research-based experiences have a positive impact on geoscience major recruitment and retention (e.g., Elkins et al., 2018). However, these types of experiences can often be hard to implement, due to cost, logistics, and the inaccessibility of research tools (Bursztyn et al., 2015). In addition, students with disabilities may not be able to fully participate in field-based activities. While Rockd as a mobile app requires that the user is able to go to physical outcrops, the Macrostrat and Rockd web portals do not. As such both are valuable tools to use for students who may not be able to physically participate in field trips (Carabajal et al., 2018), or for faculty who do not have the financial resources or logistical ability to take students into the field. In addition, active, inquiry-based learning and student research opportunities have been shown to increase student interest in STEM and to help increase the retention of tradition-ally underrepresented groups (Rissing and Cogan, 2009; Graham et al., 2013; Linn et al., 2015; Ballen et al., 2017). This is especially critical in the geosciences, which has the least diverse student population of all the STEM fields (National Center for Science and Engineering Statistics, 2015; Huntoon et al., 2018). Thus, we hope that instructors will find that incorporating Macrostrat and Rockd into their undergraduate earth science curriculum is a fruitful experience for both faculty and students that increases access to authentic earth science experiences. See Lockwood et al. (2018), for more background on the value of research experiences and active learning in geoscience education.

2 How to Use Macrostrat and Rockd

2.1 Macrostrat

Macrostrat is available as a web portal at macrostrat.org. The landing page is a dynamic geological map showing bedrock and topography. Clicking anywhere on this map will pull up information on the bedrock map unit including unit name, lithology, age, relevant literature references, and more (Fig. 1D). The search bar can be used to search for geological units, time periods, lithology, and for column view, environment (Fig. 1A). The menu bar popup allows users to add or subtract layers of data (Fig. 1B, C, D). For example, by adding *Column* view, polygons that encompass regional stratigraphic data are overlain onto the bedrock map (Fig. 1C, D). Information on these stratigraphic units can be viewed by clicking on "Match basis" names under the Macrostrat category in the information sidebar. Users can also overlay Paleobiology Database collections (*Fossils*; Fig. 1E) and satellite image data (*Satellite*).

2.2 Rockd

Rockd is an app available for download on both the iOS and Android platforms at Rockd.org (Fig. 2, Fig. 3). There is also a web-based portal for exploring Rockd checkins and data available at Rockd.org/explore (Fig. 4). When a user opens the app, the first thing they will see is a landing page that shows them geological information on their current location (Fig. 2A). This information will usually include the age of the rocks the user is standing on, the name of the geological unit, the user's current latitude and longitude, their current approximate elevation, and the lithology of the rocks the user is standing on. The accuracy and precision of the estimates vary geographically depending on the available geological map and Macrostrat column coverage. From here, the user can click on the *Map* icon (Fig. 2B), which will show the user a satellite view of their current location with geological map units superimposed on top. By clicking on any of these map units, the user can pull up more information describing the rock unit, most of which derives from the original map sources (Fig. 2C), which are also accessible by clicking on the reference. From the landing page, clicking on the *Rotate* icon ("rotate" is the name of the spherical geometric computational operation that is used to calculate a paleoposition) will produce paleogeographic map reconstructions of the user's current location at the time the rocks the user is standing on were formed or deposited (Fig. 2D). By clicking on the *Checkin* icon on the landing page, users can create their own checkin at their location by taking an image of an outcrop, adding a description, and selecting other factors such as lithology. From this page, clicking the camera icon will allow users to take a photo, or select a pre-existing photo, to associate with the checkin. Users can then provide additional information about the location, including a description of the location, a "rating" of 1 to 5 stars, and a tag for other defined attributes such as lithology, structures, fossils, minerals, and more. Once a checkin is created by a user and synced with the Rockd server, it acquires a unique URL that can be shared with anyone who has a web browser. Rockd provides a copy button that allows users to easily paste the URL into a social media post, assignment, or email.

Rockd also allows users to browse through public checkins by others that show images of rock outcrops along with additional information on locality, lithology, and other features (Fig. 3). From any page on the app, users can click

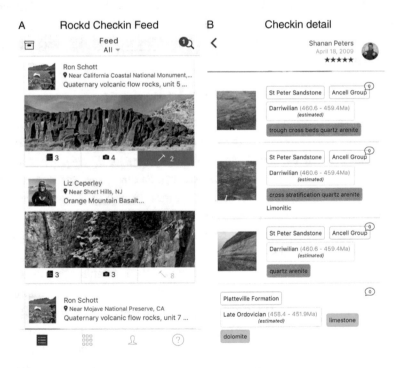

Figure 3. Example of a Rockd app checkin. A, Rockd app feed showing a list of checkins. B, Rockd app checkin detail showing additional images, lithological information, and more.

on the *feed* icon in the lower left-hand corner of the app screen (bulleted list icon), which will then pull up a list of checkins with images that users can scroll through (Fig. 3A). From here, users can click on a checkin to get more information and see additional images (Fig. 3B). Geolocated to a geological map, these checkins also provide information on the geological map unit, and can link users to additional information. One of the unique aspects of the Rockd app is that it allows interactions among users. A user can "like" another user's checkin and see who has "liked" their own checkins. Users can also comment on others' checkins, thus allowing for engagement and communication between users. Rockd also has a web-based application in which users can search and scroll through checkins (rockd.org/explore; Fig. 4).

3 Educational Integration: Examples and Ideas

As the Macrostrat web application and Rockd mobile app are relatively new tools that are still under active development, the community has not yet created established reviewed educational products. Instead, this section will provide a number of strategies for incorporating Rockd and Macrostrat into different types of earth science classes. Some of these examples are more detailed, while others are intended to serve as starting points for the reader's adoption.

3.1 A Note on Accessibility

We have tried to present a variety of activity ideas based on accessibility. Rockd mobile requires the use of a smartphone. While many students have access to these devices already, some do not, and we do not wish to highlight economic disparities amongst students. Many activities require that students be able to access local outcrops either on foot or with a vehicle. This may not always be possible, and may be limited by student resources. Likewise, Rockd checkins in the field require the use of a student's personal wireless data, though we note that images and notes can be taken in the field and checkins uploaded back on campus where free WiFi is more likely to be available. In addition, offline maps can be downloaded before embarking on a field trip. We have also included activities that require only the use of a laptop or desktop computer and access to the Internet, and do not require a smartphone or access to a personal vehicle.

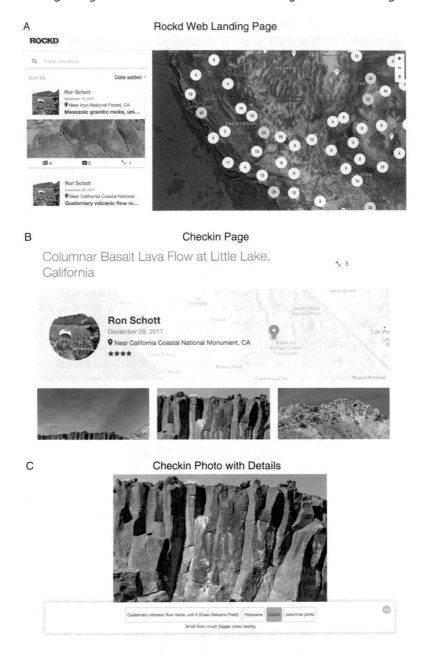

Figure 4. Rockd web application. A, Landing page showing map with checkin counts in circles and checkin feed. B, Example of checkin overview. C, Example of checkin photo with additional details.

3.2 Integration into Introductory-Level Classes

3.2.1 Fossils or Not?: Macrostrat and Rockd

This activity can be done using Macrostrat or Rockd. Using either application, students individually decide if, given the information available to them via the Rockd *Map* or Macrostrat Bedrock layer (these are equivalent datasets), there *could* be fossils where they are at that moment (on campus, on a field trip if using Rockd, etc.). Then in pairs students discuss their answers and come up with reasons for why there might or might not be fossils at their location. After discussing their predictions with their partners, they then check to see if their predictions were correct. If students are using Macrostrat, they can load the Fossils layer to see if there are fossils in their location or not. If using Rockd, students can peruse local checkins and/or view the Paleobiology Database in a mobile web browser to see if there are fossils nearby (www.paleobiodb.org). If their predictions are wrong, students can then discuss why this might be. For example, they might realize that they are standing on metamorphic rocks, and these very rarely preserve fossils. Or they might discover that they are on sedimentary rocks, but that there are no fossils. Students can then discuss issues of preservation or other factors that might lead to barren sedimentary strata. In a more advanced paleontology-focused course, students can also make predictions about what kinds of fossils they might expect to find given the age and lithology information gleaned from Macrostrat / Rockd and then check these predictions against the Paleobiology Database.

This activity can be done as a classroom or field trip group discussion with pairs sharing their predictions and findings. Alternately, students can submit short written responses in which they outline their predictions, their findings, and why those predictions did or did not match up with their findings. This activity is appropriate for any introductory-level earth science or paleontology class. Students need to have learned that fossils are most often preserved in sedimentary rocks and the basics of why this is the case. This activity helps students apply factual information to their understanding of a specific location, allows them to make and test predictions based on their understanding of course material, and gives them insight into the spatial and temporal distribution of fossils.

3.2.2 The Geohistories Project: Rockd, Macrostrat, and the Paleobiology Database

In this assignment, students write a short geological history of their hometown or another geographic place that has significance for them, which can then be posted on a class blog or website. Students use Macrostrat, Rockd, and the

Paleobiology Database in their research. The blog posts must be understandable by a general audience. The posts must discuss topics such as geology, geography, paleontology, past climate, and paleogeography. Questions students can think about include the following: What was the paleolatitude of your house 500 million years ago? Are there fossils in the area? What do these fossils tell us about the deep time history of this region? Is your town built on sedimentary, metamorphic, or igneous rocks? How does the geological history of the area affect its modern landscape or economy? Students include screen shots from Macrostrat, Rockd, and the Paleobiology Database in their blog posts.

This activity works well in an introductory-level earth science / earth history course and is currently used in Williams College's 100-level "Co-evolution of Earth and Life" course. It is best integrated into the middle or end of a semester-long course after students have learned the fundamentals of earth science. Because it deals with an off-campus location, the activity is especially well situated to straddle a break when students might be traveling home or to another location that they can use as the basis for their blog posts, including Rockd checkins or screenshots. However, this is not a requirement as students can use Macrostrat, the Rockd web portal, and the Paleobiology Database to get information about their location without traveling.

In order to begin the assignment, students must have an understanding of major rock types, how to look at / read geological maps, the basics of fossils and the fossil record, and familiarity with Macrostrat, Rockd, and the Paleobiology Database. Through this activity, students learn about the geological history of a new location based on their own research and investigations. They create a blog post that informs others about the geological history of an area, synthesize geological and paleontological data to create a cohesive narrative of a specific place, and apply general knowledge of earth science to a specific region. Students also learn to write effectively for a general audience in a way that educates and entertains.

A rough draft of the blog post is due two weeks before the final draft. Students swap posts with others in the class and read them with a rubric that focuses on clarity, avoiding jargon, and narrative. Students meet in small groups to discuss feedback and then edit their posts based on their classmates' feedback. The faculty member then reads and assesses the final versions of each student's post. Additional assignment details and a rubric for student peer assessment can be found on the SERC website.

3.2.3 Rockd Paleogeography

In this activity, students or the instructor use the <u>*Rotate*</u> function in Rockd to visualize the paleolatitude and continental configuration of their current location. Students are asked to describe how the environment of their current location differed in the past and what processes contributed to the change. For example, if the current location of the classroom is a mountainous region in the mid-latitudes of the Northern Hemisphere, but the bedrock was deposited when the location was in an equatorial shallow sea, students can discuss how sea level change, climate change, mountain building, and plate tectonics have contributed to the change. This activity has the advantage that it can be accomplished as a class, in class. The instructor can use their own device to pull up Rockd and use a dongle (VGA or HDMI adaptor) to project this image on the screen. Thus all students are accessing the same data and there is no need for individual students to have their own devices. This would work well as an in-class activity for a lecture or lab period. Alternatively, it could be modified as an assignment using the students' own devices. The activity can be integrated into any introductory earth science / earth history class that touches on the idea of plate tectonics. Students must have a basic familiarity with concepts of plate tectonics and major geological processes such as sea level change and mountain building. Alternatively, this activity could be used to introduce the idea of plate tectonics. Students learn to visualize paleocontinental reconstructions and plate tectonic motion and determine what geological processes affect the current location, elevation, etc., of a specific place. They will also apply their understanding of basic geological processes to a specific location. The instructor can request a narrative from individuals or groups, or simply poll the class for ideas after groups have had time to reflect on their work.

3.3 Integration into Upper-Level Classes and Student Research

3.3.1 Tracking the Sauk Transgression: Macrostrat

Macrostrat enables users to visualize large-scale changes in sediment deposition in North America over time. This is useful when discussing major sea level changes such as the Sauk Transgression, a large-scale rise in sea level that occurred mainly during the Cambrian Period. Using Macrostrat, students can filter columns by "marine" to show only columns that contain marine-sourced rocks. Then students can add time period filters to examine where marine packages are present that represent the Early, Middle, and Late Cambrian (make sure that students *un-click* the last time period filter before adding the next; Fig. 5). Students are asked to make observations about rock presence/absence, rock type, and inferred environments for each of these three time slices and then create a narrative that describes why these changes are occurring (i.e., as sea level rises, more marine rocks are deposited over the entirety of North America; this can be viewed as an increase in the number of shaded polygons containing marine rocks in each time slice; Fig. 5). This activity would work well in an earth history course, a sedimentary geology course, or a stratigraphy course. Students must have familiarity with Macrostrat, have an understanding of sea level change, and have familiarity with major lithological types and environments of deposition. This activity fits in well with a discussion of the Proterozoic-Cambrian transition, or when discussing major sea level changes through time. This is an easy activity to complete in small groups during a lab period in which students have access to desktop computers, or could be done as an assignment.

Through this activity, students learn that the geological record contains information about sea level change and that lithology and environment of deposition provide critical data on the history of continents. Students apply knowledge of the geological record to make inferences about sea level change through time, and they apply knowledge about rock type and paleoenvironmental indicators to make inferences about sea level change through time.

3.3.2 The "Hypothesis Test" Assignment: Rockd

This group of activities involves taking the information that Rockd feeds to users (e.g., the dashboard estimate of what a user is standing on, the map, the paleogeographic reconstructions) and asks the students to critically assess that assertion and test it as a hypothesis, rather than accepting it as fact. Students make checkins with observations to support their claims and then share the URL for those checkins with their instructor as the "write up."

Questions could include:

- What observations are consistent with the reconstructed paleolatitude for these rocks/sediments? Are there any observations that might call into question that reconstruction? For example, if the reconstructed paleolatitude is equatorial, do the sediments / fossils show any evidence for that interpretation?
- Is the Rockd estimate for the geology at this spot accurate? If not, why not? If so, how do you know? What might account for inaccuracy in the estimate? What observations could you make that would refine that estimate?

Figure 5. Example of Cambrian Sauk transgression in North America as viewed through the Macrostrat "column" view. A, Early Cambrian and marine filters applied showing minimal coverage in North America. B, Middle Cambrian and marine filters applied showing increasing coverage. C, Late Cambrian and marine filters applied showing extensive coverage.

- If Rockd provides a statement on the environment of formation, what observations are consistent with that assertion? Which are not? Are there fossils at this locality that support or contradict an environmental interpretation? For example, if the environment of formation is listed as shallow marine, are there identifiable shallow marine fossils at the locality?

This activity could work for a broad array of upper-level earth science classes such as sedimentology and stratigraphy, structural geology, paleontology, and earth history. Required student background will depend on the specifics of the activity that the instructor designs and/or the location the students visit. This activity strengthens skills and concepts including reading and interpreting maps, interpreting rock outcrops, identifying rock types in the field, identifying sedimentary structures in the field, and identifying structural features in the field. Students also develop higher-order thinking skills by critically evaluating Rockd data to determine if it is accurate based on their observations and the application of previous knowledge. This activity requires access to a variety of outcrops either locally or via fieldtrips.

3.3.3 Integration into Independent Student Research

Rockd would be an excellent complement to a field-based research project at the undergraduate level. For example, students could contribute checkins as a part of their project and include links to these checkins in their final work. Macrostrat also has a great deal of potential as a tool for primary research by students. The Macrostrat database has been utilized in many research publications (macrostrat.org/#publications). While the current web interface makes accessing primary data challenging, there is an easily accessible application programming interface (API) that allows users to download data and to build their own interactive web or data analysis applications (macrostrat.org/api; Appendix). However, this requires users to have some experience with downloading data via API and analyzing and visualizing data in programs such as R or Python. A summary of how to download data via the API can be found in Peters et al 2018; students who are familiar with these techniques or who have advisors who can assist them can access a wealth of information that allows them to answer many primary research questions.

Examples of primary research questions include:

- What are the relationships between sedimentary rock packages and extinction events through time?
- How might sedimentary rock record bias affect our understanding of the diversity of a specific taxonomic group or time period?

- How does the temporal and spatial distribution of black shales relate to global models of atmospheric oxygen?
- Are there relationships between lithological patterns (such as percent carbonate rocks versus siliciclastic rocks) and Phanerozoic marine diversity trends?

4 Student Feedback and Conclusions

As the Macrostrat web application and Rockd mobile App are relatively new tools, no comprehensive educational research has yet been done on how effective these platforms are in classroom settings. However, interviews with Williams College class of 2017 students who have used Rockd and Macrostrat in courses and after graduation indicate that these platforms provide a novel and useful complement to existing course work and fieldwork.

How, if at all, did the use of the Rockd app change your experience of a field trip or other in-class experience?

> I think one of the most powerful aspects of the Rockd app is, perhaps, the same as any geologic map: It makes the invisible visible. It helps us see not only what's under our feet but also how it fits into the wider geologic setting around us. The ability to zoom out and explore the region around you helps to connect both what you're seeing in front of you (the outcrop, the fossil, or just a hilltop) with the wider geologic context of the surrounding area. Also, I think the community aspect of the Rockd app is quite fun. It reminds us that we're not alone . . . .. And, to see outcrops from around the world helps to underscore the similarities and differences that exist in seemingly simple outcrops.

How, if at all, did the use of the Rockd app change your experience of a non-academic experience? (i.e., use of the app outside of a classroom setting)?

> I'm currently not working in a field related to geosciences. But the Rockd app keeps me connected to a discipline I'm passionate about and a community I care about. I don't open the Rockd app often these days, but whenever I do, it makes me smile, remember, and think about why geology matters. It's a reminder, sitting on the home screen of my phone, that there's a world underneath my feet and a history that predates me. And it brings back fond memories of my geosciences courses!
>
> I've been to a bunch of different biological environments this year while researching bats, and it's really fun to take a moment to check what the geologic history of wherever I'm standing is. It's actually helped me widen my perspective on the places I've been to since it grounds me (no pun intended) to the deeper geologic narrative of a site rather than focusing just

on the current biological/political climate of a village or country. It's fun to see I'm standing on rock I've never stood on before! My favorite feature of it is actually the map, since I like zooming out and seeing the geologic landscape of the country I'm in. Plus, it's fun to show other people! I didn't think I'd use RockD beyond the field trip I downloaded it for last year, but it's been a definite source of fun this year for me. I like to whip it out when I'm on walks with friends or am in the middle of a new city. It makes me feel like a very #official scientist, ha ha!

The use of large online scientific databases in research is growing, and as such, using these resources in our teaching helps better prepare our students and teaches them valuable skills in data analysis, vetting, and interpretation. In addition, the use of mobile technologies is now interwoven into the lives of the vast majority of all people, regardless of age, geographic location, or level of education. Thus, integrating mobile scientific apps into earth science education provides access to geological data to a much broader swath of the public than previously possible, and anecdotal evidence shows that it also helps students integrate earth science into their lives after their formal education ends. Macrostrat and Rockd present excellent opportunities to integrate primary data and mobile learning into the earth sciences curriculum, including paleontology and earth history courses.

References

Ballen, C. J., Wieman, C., Salehi, S., Searle, J. B. & Zamudio, K. R. (2017). Enhancing diversity in undergraduate science: Self-efficacy drives performance gains with active learning. *Cell Biology Education*, 16, 1–6.

Bursztyn, N. & Pederson, J. (2015). Utilizing geo-referenced mobile game technology for universally accessible virtual geology field trips. *International Journal of Education in Mathematics, Science and Technology*, 3, 93–100.

Carabajal, I. G., Marshall, A. M. & Atchison, C. L. (2018). A synthesis of instructional strategies in geoscience education literature that address barriers to inclusion for students with disabilities. *Journal of Geoscience Education*, 65, 531–541.

Dennen, V. P. & Hao, S. (2014). Intentionally mobile pedagogy: the M-COPE framework for mobile learning in higher education. *Technology, Pedagogy and Education*, 23, 397–419.

Elkins, J. T. (2009). Using portable media players (iPod) to support electronic course materials during a field-based introductory geology course. *Journal of Geoscience Education*, 57, 106–112.

Elkins, J. T. & Elkins, N. M. L. (2018). Teaching geology in the field: Significant geoscience concept gains in entirely field-based introductory geology courses. *Journal of Geoscience Education*, 55, 126–132.

Finnegan, S., Heim, N. A., Peters, S. E. & Fischer, W. W. (2012). Climate change and the selective signature of the Late Ordovician mass extinction. *Proceedings of the National Academy of Sciences of the United States of America*, 109, 6829–6834.

Gikas, J. & Grant, M. M. (2013). Mobile computing devices in higher education: Student perspectives on learning with cellphones, smartphones & social media. *The Internet and Higher Education*, 19, 18–26.

Graham, M. J., Frederick, J., Byars-Winston, A., Hunter, A. B. & Handelsman, J. (2013). Increasing persistence of college students in STEM. *Science*, 341, 1455–1456.

Huntoon, J. E. & Lane, M. J. (2007). Diversity in the geosciences and successful strategies for increasing diversity. *Journal of Geoscience Education*, 55, 447–457.

Husson, J. M. & Peters, S. E. (2017). Atmospheric oxygenation driven by unsteady growth of the continental sedimentary reservoir. *Earth and Planetary Science Letters*, 460, 68–75.

Linn, M. C., Palmer, E., Baranger, A., Gérard, E. & Stone, E. (2015). Undergraduate research experiences: Impacts and opportunities. *Science*, 347, 627.

McGreen, N. & Arnedillo Sánchez, I. (2005). Mapping challenge: A case study in the use of mobile phones in collaborative, contextual learning. *Mobile Learning*, 213–217.

National Center for Science and Engineering Statistics. (2015). *Women, Minorities, and Persons with Disabilities in Science and Engineering: 2015, Spec. Rep. NSF 15–311*. National Science Foundation, Arlington, VA.

Nelsen, M. P., DiMichele, W. A., Peters, S. E. & Boyce, C. K. (2016). Delayed fungal evolution did not cause the Paleozoic peak in coal production. Proceedings of the National *Academy* of Sciences of the United States of America, 113, 2442–2447.

Peters, S. E. & Gaines, R. R. (2012). Formation of the "Great Unconformity" as a trigger for the Cambrian explosion. *Nature*, 484, 363–366.

Peters, S. E., Husson, J. M. & Czaplewski J. (2018). Macrostrat: A Platform for Geological Data Integration and Deep-Time Earth Crust Research. *Geochem. Geophys. Geosyst.*, 19, 1393–1409.

Rissing, S. W. & Cogan, J. G. (2009). Can an inquiry approach improve college student learning in a teaching laboratory? *Cell Biology Education*, 8, 55–61.

Styers, D. M. (2018). Using big data to engage undergraduate students in authentic science, *Journal of Geoscience Education*, 66, 12–24.

Taleb, Z. & Sohrabi, A. (2012). Learning on the move: The use of mobile technology to support learning for university students. *Procedia – Social and Behavioral Sciences*, 69, 1102–1109.

Cambridge Elements ☰

Elements of Paleontology

Editor-in-Chief

Colin D. Sumrall
University of Tennessee

About the Series

The Elements of Paleontology series is a publishing collaboration between the Paleontological Society and Cambridge University Press. The series covers the full spectrum of topics in paleontology and paleobiology, and related topics in the Earth and life sciences of interest to students and researchers of paleontology.

The Paleontological Society is an international nonprofit organization devoted exclusively to the science of paleontology: invertebrate and vertebrate paleontology, micropaleontology, and paleobotany. The Society's mission is to advance the study of the fossil record through scientific research, education, and advocacy. Its vision is to be a leading global advocate for understanding life's history and evolution. The Society has several membership categories, including regular, amateur/avocational, student, and retired. Members, representing some 40 countries, include professional paleontologists, academicians, science editors, Earth science teachers, museum specialists, undergraduate and graduate students, postdoctoral scholars, and amateur/avocational paleontologists.

Paleontological
S O C I E T Y

Cambridge Elements ☰

Elements of Paleontology

Elements in the Series

These Elements are contributions to the Paleontological Short Course on *Pedagogy and Technology in the Modern Paleontology Classroom* (organized by Phoebe Cohen, Rowan Lockwood, and Lisa Boush), convened at the Geological Society of America Annual Meeting in November 2018 (Indianapolis, Indiana USA).

A full series listing is available at: www.cambridge.org/EPLY

Printed in the United States
By Bookmasters